# Everyday Wonders

*Ideas of the Past that We Use Today*

by Solveig Paulson Russell

pictures by Robert Frankenberg

*Parents' Magazine Press • New York*

Library of Congress Cataloging in Publication Data
Russell, Solveig Paulson
  Everyday wonders.

  (Finding-out books)

  SUMMARY: Relates several cases in which observa-
tions, imagination, accidents, and a desire to make work
easier have resulted in important inventions.
    1. Inventions—Juvenile literature.  [1. Inventions]
I. Frankenberg, Robert C., illus.  II. Title.
T15.R8      608'.7      72-8841
ISBN 0-8193-0660-6 (lib bdg)

# Contents

# Chapter One
# Everyday Wonders

Every day in most of our homes people push buttons, turn switches, connect cords, run machines—and wonderful things happen. With a click of a switch we get light, or heat, or coolness. We hear music, or see pictures. We make a vacuum cleaner pick up dirt. We send garbage to be chewed up in a garbage disposal unit and then washed away.

In some buildings we push a button and a door opens so that an elevator can move upward or downward to take us to another floor. We slip a coin into a slot and out comes food, or soda pop, postage stamps, or a tune.

Outside we ride in cars or buses, trains or planes that move us in comfort to new places.

When we are hurt, an x-ray machine makes it
possible for doctors to see what happened inside
our bodies. There are also many medicines and
treatments to make us well.

We are so used to these wonders that we think little of them, but this was not always so. Each one of the work-savers and modern conveniences was first an idea that grew to make life easier for us. They are ours only because people of different lands worked hard to make them possible. It would take hundreds of books to tell about all the inventions and discoveries used in our world. In this book we will mention only a few that we use very often.

## Chapter Two
# Beginnings

The cave men lived a long time, knowing very little. But as time wore on some of the people gradually got ideas of how to make life better. They tamed animals and raised food. Nobody now knows just when wheels were made or when levers were first used to help lift heavy things. But somebody had to think of them before they could be used.

The ancient Egyptians and Greeks came after
the earliest people. They had many ideas that added
to the comfort of living. They learned how to
understand the relationship of numbers, and
invented simple machines such as the screw and
the pulley.

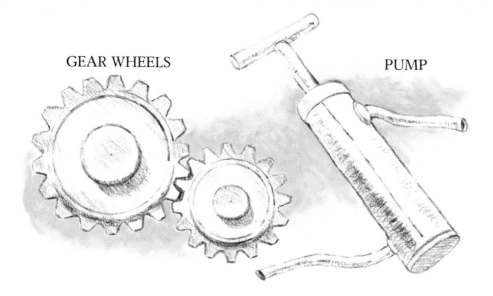

GEAR WHEELS

PUMP

One Greek, named Hero, 120 years before Christ was born, invented ways to make steam work. He made a syphon, a gearwheel, a pump, and a clock that ran by water. Though Hero used his inventions for toys only, the ideas for them later grew into the beginnings of machines used today.

WATER CLOCKS

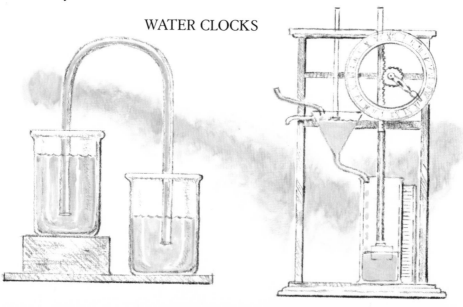

The Romans, who followed the Greeks, were
good engineers. They built roads, bridges, and
buildings in ways that are still used.

After the first simple machines were made, there came a time known as the Middle Ages. These lasted about eight hundred years. Then little progress was made with inventions. This was because people were spending their time fighting. Different groups wanted land and power. But when these years had passed, people again became interested in improving ways of living.

One man of these times stands out because of his work. He was Leonardo da Vinci, an Italian, who many people feel was one of the greatest thinkers of all time. Da Vinci was an artist. He was also an inventor. He and his followers invented many mechanical things and stirred others to think of still more inventions.

After Leonardo da Vinci, thousands of young men became interested in mechanics, and there were many inventions and improvements in machines. In the last part of the seventeenth century, one of the great men of science gave the world some giant-sized ideas. The man was Sir Isaac Newton.

By the time he was twenty-one years old, Newton had produced three of the greatest discoveries ever known. One of these is called calculus. It is a system for working out difficult problems in mathematics. A second discovery was about the nature of light, its colors, and the way it travels. The third discovery was the law of how

things move—the law of gravity. Newton
explained how gravity works. He could not explain
its cause, but when he saw an apple fall from a tree
he wondered why things always fell downward.
From his thinking came our better understanding
of gravity. Isaac Newton opened the way for
modern invention.

# Chapter Three
# Mechanical Inventions

*The Loom*

The first looms were made of yarns, or threads, tied side by side to a piece of wood. Other yarns were woven in and out between them by hand. Then someone thought of using a foot pedal to raise every other long thread so the cross threads could be put in faster. By the year 1700 many people, especially in England, were busy making cloth on foot-pedal looms.

With so many people weaving, faster ways were needed for spinning threads for the looms. The old spinning wheels made only one thread at a time. James Hargreaves thought up a faster way. In 1767 he made a spinning wheel that turned with a crank and spun eight threads at a time. He called his invention a "spinning jenny" in honor of his wife, whose name was Jenny. A few years later, improved spinning machines were made to run by water power.

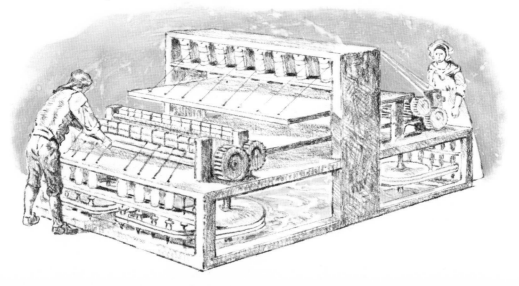

As more thread or yarn was made, looms needed improvements, too. In 1785 Edmund Cartwright, also an Englishman, worked out an idea for a loom that could be run by a steam engine. From Cartwright's work, and that of others, came our powerful looms of today. Now looms are huge machines that fill buildings. They use threads of many kinds from all over the world. They turn out billions of yards of many different kinds of useful fabrics that are made into things as small as handkerchiefs or as large as rugs.

*The Cotton Gin*

As ways of weaving were improved, the need for cotton to weave into cloth increased. In the United States the South raised cotton, but it took a long time to get the seeds out of it by hand. The problem of getting the seeds out faster so more cotton could be produced was solved by Eli Whitney.

Whitney had a quick mind and he loved to invent things. When he was just a boy, he invented a new way to make nails. He made a violin, and he also made simple tools for his neighbors. Later, as a young man, he made a machine that was called a cotton gin—short for cotton engine.

He made a cylinder, or round part, with wire teeth in it. He put the cylinder in a wooden frame so that, when the cylinder turned, the teeth could pass through slits in a comblike sheet of metal. As the cylinder turned, the teeth grabbed bits of fluffy cotton and pulled them through the narrow

slits, leaving the seeds behind. This simple idea is
still used in the cotton engines of today, but now
the wheels of the big machines turn much faster —
spinning around to clean large amounts of cotton
in little time.

Eli Whitney's invention in 1795 helped to make the southern states rich. Soon there was enough cotton raised there to ship some to England as well as to the weavers in New England. Within seven years after Whitney's invention, the cotton crop had increased by more than twenty times.

*The Sewing Machine*

When the earliest people wanted to join two pieces of material, such as animal skins, they made holes in the pieces with a sharp bone. Then they threaded thongs of some kind through the holes and tied the pieces together.

As time went on, people made needles, with holes in them for threading, from slender pieces of bone. After these bone needles came needles of steel, which were used for hundreds of years. In fact, steel needles, used for making stitches one by one, by hand, were used up to our great grand-mothers' time. They still are used for hand sewing, and even today many garments are made by hand in homes all over the world. But clothing is sewn in large factories by machines that stitch with amazing speed.

Men in France, England, and America worked for years to think up a way to make a sewing machine; and some did make machines that could sew. Among these was an Englishman, Thomas Saint, who, in 1780, made the first known sewing machine. But his machine had only one thread and made a chain stitch. When the thread was broken, all the stitches pulled out.

Walter Hunt, an American, made a two-thread lockstitch machine about 1832. But he thought the machine would put out of work a lot of people who did hand sewing for a living. So he put his machine aside.

Elias Howe gets the credit for making the sewing machine we know. When he was a young man, he worked in a machine shop. There he heard someone say, "Anyone who can make a sewing machine will be rich."

"I'll do it!" Elias said to himself.

He was poor and had to support his family. So it took five years of his spare time, and many trials, before he had his machine made. When it was done, in 1846, he raced his machine against five seamstresses. His machine sewed five times as fast as the women sewed by hand.

An important machine for sewing shoes was
made by a young black man, Jan Earnst Matzeliger.
He worked in a shoe factory where the upper parts
of shoes had to be sewn to the soles by hand.
Matzeliger decided to make a machine that would
do this work. He worked for years, and finally,
in 1883, he had a sewing machine that could sew
a whole shoe. The young black man's invention
made it possible to make shoes much faster. Soon
it was used all over the world.

*The Steam Engine*

If you have ever watched a tea kettle boiling, you have seen the little cloud over the spout as the water inside the kettle changes to steam. Maybe you have seen lids on kettles rise up because the steam inside, from the boiling water, pushed up against the lids. Steam, pressing against something that bars its way, has power. This power can be put to work by steam engines.

The steam engine was the result of the thinking of many men, each one building on the work of others. As we have said, way back in ancient Greece, the man named Hero had made a little steam engine for a toy. Several men tried to build on the idea, but for some time their inventions, also, were little more than toys.

In 1690 Denis Papin of France made the first steam engine having parts called a cylinder and a piston. Thomas Newcomen, an Englishman, improved on Papin's work. But it took James

Watt, a Scot, to iron out many of the problems
of these first machines.

James Watt worked for a long time before he
finally got the right ideas about making cylinders
and pistons work well. He finished his first engine
in 1765. Four years later he had an improved
steam engine.

Before Watt's engine came into being, windmills
and water wheels did such tasks as grinding corn
and wheat into meal, and most tools were worked
by hand. But with Watt's steam engine the world
of industry took giant steps forward, for it had
an engine, harnessed to steam, to do the work.

*The Harvester, or Reaper*

In the 1830s there was not enough grain raised
to meet the needs of the people of the United
States. Some wheat and other grains had to come
from foreign countries. There was plenty of good
land for growing grain. The trouble was that
harvesting, or reaping, took too long. Grain had
to be harvested during the few days before the ripe
grain heads dropped their seeds, so the harvesting
time couldn't last very long. There was no use

having more grain fields than could be worked
in a short time.

Harvesting was done by hand for many years.
Then in England a man named Patrick Bell made
a reaping machine that was almost a big success.
But grain that wasn't straight, or too many weeds in
the field, made the cutters of the machine choke up,
so they wouldn't cut. This machine was used,
but the grain that the machine didn't cut had to
be harvested by hand.

Then a reaper, or harvester, was made in
America in 1834 by Cyrus H. McCormick. It

took away the hard, back-breaking work of hand harvesting and that made it possible to raise more wheat. After this first reaper, McCormick continued to work out new ideas for harvesting, and in 1848 he had a factory that could turn out many machines. He sold his reapers all over the world.

From the reapers made by Cyrus McCormick have come the huge combine machines that now harvest our grain fields. Today these machines move over large fields and cut, clean, and sack

the grain in record time. The farmers who once harvested slowly by hand would never have believed such speed could be possible.

## Oil Cups for Machines

If any machinery is run without oil, its parts finally wear away. That is why automobiles and all other machinery must be oiled.

It used to be that machines had to stop running when oil was put into them, but a black man, Elijah McCoy, put an end to this in 1872.

McCoy had learned about engineering in Scotland, where he was born. In the United States he became a fireman for a railroad company. One of his duties was to oil the engine. McCoy thought it was a waste of time to have to stop the engine in order to oil it. So he set to work on the problem. In 1872 he had made a cuplike container that held oil and let it run into the parts of a machine as the machine was being run. Later he made

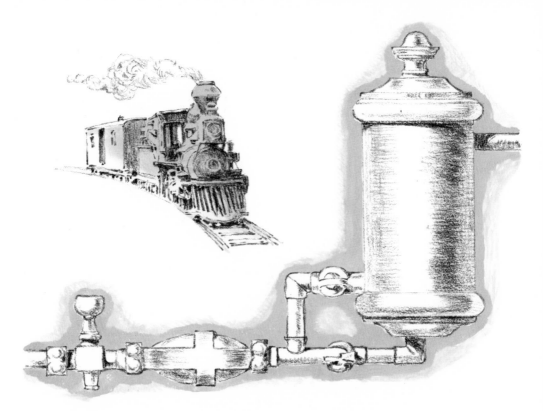

improvements on this idea. Soon all machine
factories were using McCoy's oil cups.

A word for oiling is lubrication. Elijah McCoy is
often spoken of as "the father of lubrication."
It is said that when men were about to buy heavy
machinery they looked at each of the oil cups and
asked, "Is it the real McCoy?" This question is
often asked today when people want to knowing
if something is real and true.

## Chapter Four
# Electrical and Combustion Inventions

*Electricity*

We use electric things often—electric lights, electric motors, toasters, heaters, irons, and other tools that depend on electricity to make them useful. We live in an electric age. All electric tools have been invented, but electricity itself is a strange natural force that, though not entirely understood, can be made to work.

About twenty-five hundred years ago a Greek, named Thales, rubbed a piece of amber with a cloth. He was surprised when, after rubbing, the amber had power to draw light objects toward itself. If you have seen a magnet work, you know how the piece of amber worked. You can rub something made of plastic—a pen, for instance—

against a cloth for a few moments and it will pull tiny shreds of cloth toward itself. The pulling power is electricity.

Nobody paid much attention to Thales's discovery until about the year 1600. Then people found that sparks sometimes jumped when electricity was produced by rubbing. Sparks came when a cat's fur was rubbed briskly. When people combed their own hair, the ends sometimes crackled and lifted toward the comb. Sometimes combed hair even gave off sparks.

Someone made a large glass disk which could be
spun between brushes. When it was spun, long
sparks leaped and flew about. People liked to watch
the sparks of this "spark machine."

Then one day in 1745 a man in Holland, Pieter van Musschenbroek, fixed up a special kind of water-filled jar with an iron spike forced through its cork to touch the water. He connected the jar to a spark machine. He spun the machine for a few moments, and when he touched the jar he was knocked down by the force of the electricity in it. He had discovered a way to store up electricity in a container so that greater amounts of power could be used.

Another man to experiment with electricity was
Benjamin Franklin. In 1752 he proved that
lightning was electricity. During a lightning storm
he flew a kite, with a key attached to the string.
The lightning struck the kite and followed down
the rain-soaked string to the key, giving Franklin's
knuckles a shock. This led to the making of

lightning rods. Lightning rods are put up over the roofs of houses. When lightning strikes, it hits the rod instead of the house and is carried down into the ground where it does no harm.

About 1799 an invention that helped others to do more with electricity was made by an Italian named Alessandro Volta. He found that electricity could be made in other ways besides rubbing—by using acid with copper and zinc plates. Volta made the first battery.

Michael Faraday, in England, wanted to find a way to make electricity cheaply. He worked for years and tried many experiments. Finally, in 1831, he built a machine of wires and magnets. It had a hand crank. As long as the crank was turned, the machine made an electric current.

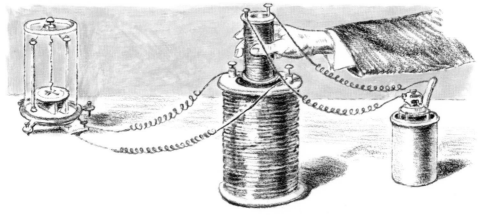

In America, another inventor, Joseph Henry, was at work on the same problem of making cheap electricity. Neither he nor Faraday knew of the other's work, but both had the same ideas. From their work came the first electric generators, and the knowledge of how to use the cheap power we have today.

Generators made electricity possible for machines all over the world. Steam engines that had to be tended constantly and fed with fuel to keep the steam coming were replaced with electric motors.

Electric motors are important servants in our world. Made in almost endless sizes they turn machinery in mills and factories. They move cars and locomotives. They help keep us clean and comfortable through the use of vacuum cleaners,

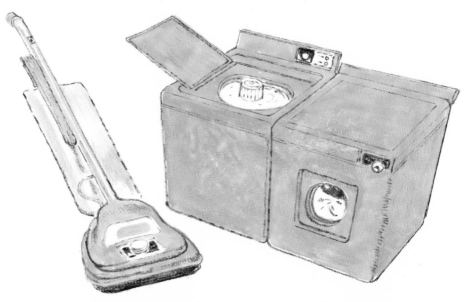

washing machines, fans, dryers, and kitchen mixers. If you try to list all the ways electric motors are used today, you will have a long list.

## The Telegraph

Samuel Morse found out that electricity can flow through a wire very quickly and that gave him the idea that electricity might be used to send messages.

Morse, an American, worked for a long time on this idea without much success. Then he heard of Joseph Henry's work. Henry had strung out a mile of wire and made a bell ring instantly at the end of it, when electricity passed through the wire. Henry's work gave Morse much help in making his telegraph.

He built new models and used electricity, not to ring a bell, but to make clicking sounds. Then he worked out an alphabet code, calling the clicks dots and dashes—making a dot and dash combination for every letter of the alphabet.

On May 24, 1844 Morse sent the first telegraph message in the world over a forty-mile wire. The message was: "What hath God wrought" spelled out in dots and dashes.

After this, telegraph wires were strung in many places throughout the world. In 1866, after several attempts, a cable carrying telegraph wires was laid under the Atlantic Ocean by Cyrus W. Field. Now many cables cross the oceans.

## The International Morse Code

### ALPHABET

| | | | |
|---|---|---|---|
| A. •- | H. •••• | O. --- | V. •••- |
| B. -••• | I. •• | P. •--• | W. •-- |
| C. -•-• | J. •--- | Q. --•- | X. -••- |
| D. -•• | K. -•- | R. •-• | Y. -•-- |
| E. • | L. •-•• | S. ••• | Z. --•• |
| F. ••-• | M. -- | T. - | |
| G. --• | N. -• | U. ••- | |

Today, to send a message anywhere in the world, anyone may telephone a telegraph office and say: "I want to send a telegram." And in a very short time the message will be delivered.

## Electric Light

It would be hard to say which use of electricity helps us most today. But many people believe it is electric light.

Inventors had been trying to make electric light for years, and a few had some success. Charles F. Brush made a successful carbon "arc" light. Arc lights were good for street lighting, but were too glaring for home use. They went out if the

current was cut down. Thomas Alva Edison gets the credit for most of the electric lights we use today.

Edison knew that electricity heats the wire it passes through. But when wires are heated enough to make them glow, they burn away. The big problem was to find material that would not burn away quickly when electricity passed through it. Edison and his helpers hunted all through the world for years for just the right materials to use to make electric lights burn a long time. He found the material when he tried carbonized threads from bamboo plants. They burned brightly in a glass bulb from which air had been removed. By 1880, Edison had a good long-burning lamp to give to the world.

Since the first successful electric lights, improvements have been made in materials, shapes, and uses. One was the replacing of bamboo with tungsten wire. Today our homes, buildings, and outdoor areas glow with light at night—light that has come from the faithful work of Thomas Edison and others like him.

## The Telephone

We say that Alexander Graham Bell, an American born in Scotland, invented the telephone. Others throughout the world had had the same idea for a telephone, but Bell discovered his way of making one without knowing of the work of others. He was the first to offer a telephone for public use.

Bell knew that a telegraph could send messages through a wire, so he wondered if sounds couldn't be sent also. He began to work on this idea. Though often discouraged, he kept on working, and one March day in 1876 he sat in a room making some changes. His helper, Thomas A. Watson, was working in another room. A wire connected the instruments of the two men.

Bell, by accident, upset a bottle of strong acid, which spilled on his clothes. At once he cried out, "Mr. Watson, come here, I want you!" Watson could not have heard Bell's voice through the walls of the rooms, but he came bursting into the room where Bell was and shouted, "Mr. Bell, I heard every word you said—distinctly!"

Bell's telephone was born, but much work still had to be done to improve it. For a long time, people using phones had to ring for an operator, who would call the number wanted. Today most calls are made with dial or push-button phones.

*Radio and Television*

Did you ever wonder how it is that just by turning a radio knob sounds of many kinds can be brought into a room? Radio waves are everywhere in the air. We cannot see or feel them, but we know they exist. They are not electricity, but electricity brings them into use. Sometimes radio waves are called electro-magnetic waves.

Guglielmo Marconi, an Italian, is most often called the inventor of radio. But before Marconi was born, scientists had known about electro-magnetic waves. A number of people had worked with them. Marconi had the idea that radio waves could be used to send messages. He worked on this idea until he succeeded in sending signals on the waves.

Then, in 1904, John A. Fleming, an Englishman,

invented a tube to catch radio waves. In 1906 Lee De Forest, an American, invented an amplifier to make sound stronger. With these two inventions, music and voices could be sent over long distances by radio. Other scientists added their work and, by 1920, radio stations were built so that people everywhere could have the magic of radio.

The inventions also made radar, loud speakers, sound in movies, calculating machines, and many other inventions possible.

Television, too, is the work of many people. In 1884, Paul Nipkow in Germany tried to find a way to send pictures over wires. In 1927 and 1930, television was tried in England and in the United

States. But it wasn't successful until Vladimir Zworykin, a Russian-American, invented a special vacuum tube for sending and receiving pictures, and Philo Farnsworth, an American, designed a TV camera and receiver. Along with other improvements, these inventions gave people modern TV sets. Now music and pictured information reaches everywhere. Special satellites above the earth bounce pictures to the world, showing events only a few moments after they have happened.

*The Automobile*

People in the past have traveled by foot, dog teams, camels, horses, boats, trains, submarines, balloons, and other ways. Today, most people do at least part of their traveling by automobile. The work of many men has made this possible.

The first automobiles were run by steam. They were slow and hard to steer. They had big boilers for the steam, and had to carry fuel to burn to make the steam.

Then ways to make engines driven by gasoline were discovered, and automobiles with gasoline engines were made. At first they were like buggies and were called "horseless carriages." They were noisy and smelly. When they broke down, people watching laughed and called out, "Get a horse!"

As time passed, men worked out great improvements in automobiles. Among the early automobile makers were R. E. Olds, David Buick, Elwood Haynes, Alexander Winton, and Henry

Ford. These men used and improved the work of
others before them.

One important improvement was the way parts
of automobiles were put together. At first
automobiles were made one at a time. Henry Ford
saw that it would be cheaper and quicker to make a
lot of the same parts at the same time. After the
different parts were made they were placed along a
line called the assembly line. The bare frame of a

car was moved along the line and as it slowly reached a man, he would put on the first part. Then the frame moved to a second man who added another part. Then a third, fourth, and more men added parts until, at the end of the line, the car was finished. Today assembly lines are used for many things besides automobiles.

Improvements are still going on. Each year cars are made safer. And people keep working to make

better cars that will not pollute the air.

It is said that we are a nation on wheels. This is true, for each day automobiles move millions to school, work, shopping places, and vacation lands. There are so many automobiles that they are a problem. They fill the streets, send bad fumes into the air, and often cause accidents. Although much improvement is being made, there is still much work to be done to solve the problems brought about by automobiles and gasoline engines.

*Airplanes*

While some inventors were busy with automobile ideas, others were thinking of flying. People have always wanted to fly. From time to time, daring men even made pairs of birdlike wings and jumped from high places, but the wings never worked.

In 1812, Sir George Cayley of England, while watching birds, saw that rushing air was needed to lift birds off the ground. The bird's wings, moving very fast, made small winds. He knew that, for anything to fly, moving air would be needed to make air currents. He worked out ideas that later helped others to fly, but Cayley himself never flew.

The same was true of Dr. Samuel Langley and others who worked on aircraft and plans for flying. These men thought of ways to shape wings and different parts of aircraft so that planes could be held up in the air better. They worked on ideas for engines with flight power.

Then in 1900 the brothers Orville and Wilbur Wright, who had owned a bicycle shop, began to work with gliders. They made several models and finally put a gasoline engine in one. On December 17, 1903 they actually flew the first heavier-than-air machine. They flew only a short way, but their flight was the beginning of more and better flights.

After the use of planes in the First World War, people had more confidence in flying. When

Charles Lindbergh flew alone over the Atlantic Ocean, and Richard E. Byrd became the first man to fly over the North and South Poles, people came to believe that air travel was possible. Planes soon were used for mail service and for passengers.

Now large planes travel very fast everywhere, saving time and bringing people together. Now men are sometimes shot into space with rockets, and fly off to great adventure.

But planes, too, pollute the air and make a lot of noise, so even people who fly often do not like to live near airports.

## Chapter Five
# Our Challenging Future

We have sampled only a very few of the ideas that
have grown from the observations, imagination,
and work of people in the past. Men and women
today have the same qualities that led to the
discoveries and inventions we have discussed. In
our laboratories scientists of every race work long

hours, often with disappointments, to make something new or improve something old.

Not all the ideas scientists and inventors work at are for big things like airplanes. Many are looking for better ways to make little everyday things—zippers, pliers, meat cutters, jigsaw puzzles, hub caps, candles, pots and pans.

But each day, also, ideas are used for big, important things that once were never dreamed of. Space travel is one of these ideas. In 1969, the first men walked on the moon. In 1971, they traveled over the moon in a vehicle built by scientists who had learned from the experiences of other moon studies. Now people are sending spacecraft beyond the moon to learn about the planets.

In medicine, too, new ideas are helping people to better health. Many doctors and scientists are at work looking for treatment to cure diseases.

But some of our present inventions make problems. We have already talked about pollution. Overcrowding and destroying the wealth of nature are others. In some of our big cities, new ideas are being studied and tried out for getting around, instead of using automobiles. One such idea is trains with electric motors. The motors force under the trains streams of air that are strong enough to hold the trains up. A rail on the ground guides the train as it rides on a stream of air. Electric motors do not pollute the air we breathe.

New ideas keep on growing. Ideas of long ago led to newer and better ways of living. These ideas were developed or improved by men and women from every country. As people learn more about the wonders of the world we live in, we can expect many ideas in the future, all built on the ideas of the past.

# Index

motors, electric, 40, 62
moon travel, 61
Musschenbroek, Pieter van, 36

nails, Whitney and, 18
needles, 21
Newcomen, Thomas, 25
Newton, Sir Isaac, 13-14
Nipkow, Paul, 49
numbers, relationship of, 9

oil cups, 31-32
oiling, of machines, 31-32
Olds, R. E., 52

Papin, Denis, 25
planets, travel to, 61
pollution, 55, 58, 62
pulley, 9
pump, 10

radar, 49
radio, 48-49
radio waves, 48
reaper, see harvester
receiver, for television, 50
roads, Roman, 11
Romans, 11
Russia, inventor from, 50

Saint, Thomas, 22
satellites, for television, 50
scientists, 59-60, 62
Scotland, inventions from, 26, 31, 46
screw, 9
sewing machine, 20-24
shoes, sewing of, 24
sound, in movies, 49
space travel, 58, 61
spark machine, 35
sparks, 34
spinning jenny, 16

steam engine, 25-27;
    for looms, 17;
    replacement of, 40;
    for automobiles, 51
syphon, 10

telegraph, 41-43, 46
telephone, 46-47
television, 49-50
Thales, 33, 34
thread, 15-17
tools, Whitney and, 18;
    and steam engines, 27
trains, with electric motor, 62
tube, radio, 48-49

United States, cotton in, 18, 20;
    discoveries and inventions from,
        18, 22, 23, 24, 29, 31, 37, 39,
        41, 42-43, 44, 46, 49, 50,
        56-58;
    and grain, 28

vacuum tube, for television, 50
violin, Whitney and, 18
Volta, Alessandro, 38

water power, 16
water wheels, 27
Watson, Thomas A., 46, 47
Watt, James, 25-27
wheels, 8
Whitney, Eli, 18-20
windmills, 27
Winton, Alexander, 52
Wright, Orville and Wilbur, 57

x-ray machine, 6

yarn, see thread

Zworykin, Vladimir, 50